Little Scientists
**BIG** Questions?

# How Does a Crane Driver Use Science?

Written and Designed
by **Alix Wood**

# It's fun to watch what's happening at the building site.

The buildings are growing **taller**. . .

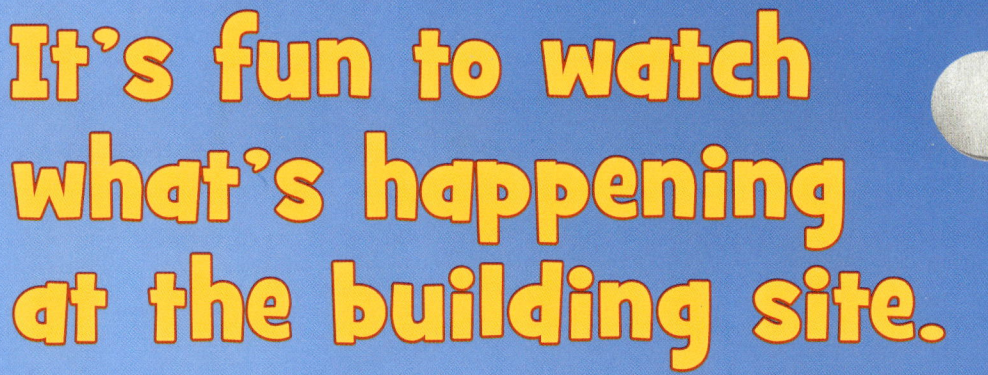

A crane driver, or operator,
controls the crane.

Crane driver

They use joysticks, levers, pedals and
computers to make the crane
do different jobs.

# To operate a crane safely, the driver must use **science**.

A crane building a wind turbine

How does a crane driver use science?

Come on little scientists, let's answer that BIG question!

# There are lots of different types of cranes.

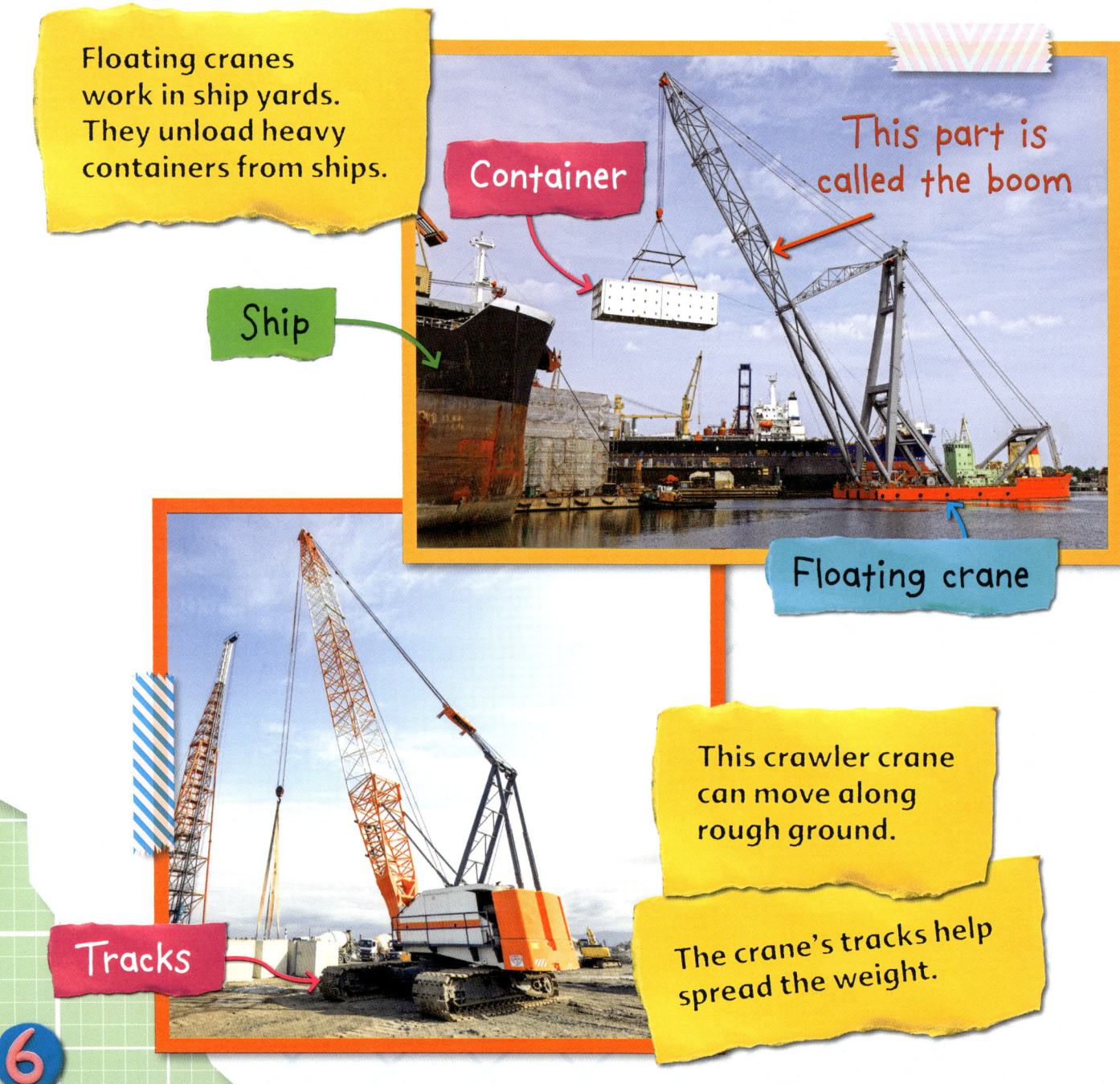

Floating cranes work in ship yards. They unload heavy containers from ships.

Container

This part is called the boom

Ship

Floating crane

This crawler crane can move along rough ground.

Tracks

The crane's tracks help spread the weight.

Some trucks have cranes on the back.

The crane uses power from the truck to help lift the load.

Crane

Leg

Leg

The truck has pull-out legs that stabilise it. The legs stop the truck tipping over!

Let's say it! "STAY-buh-lize"

7

**A crane uses forces to lift heavy things.**

**Get ready for some BIG science!**

Cables

A crane has pulleys, which are wheels with cables wrapped around them.

Pulley wheel

Hook

A pulley is a simple machine.

8

# A pulley changes the direction of a pull force and makes it more powerful.

Pulley wheel

I can't lift it.

Now I can.

Pull force

Heavy load

Heavy load

Pull force

Pulling down on a rope is much easier than pulling up.

9

# Cranes lift very heavy things.
# What stops them toppling over?

A crane works a little like a seesaw.

If the weight is only on one side, a seesaw or crane will tip.

Crane drivers put heavy weights on the back of the crane.

They use maths to work out how much weight they need.

Boom

Weights

Turntable

A crane's turntable lets the crane swivel, or turn, in different directions.

Now the crane won't tip over!

11

# Long booms, short booms, high booms, low booms!

How does a crane driver decide how long they want the crane boom to be? They use science!

A crane driver knows a shorter boom can carry a heavier weight than a long boom.

Short boom

But a long boom can reach further.

Long boom

Heavy load

Lighter load

To make a boom EVEN longer, a crane driver attaches a jib.

Jib

Boom

A jib gives the boom more height or allows it to reach further.

13

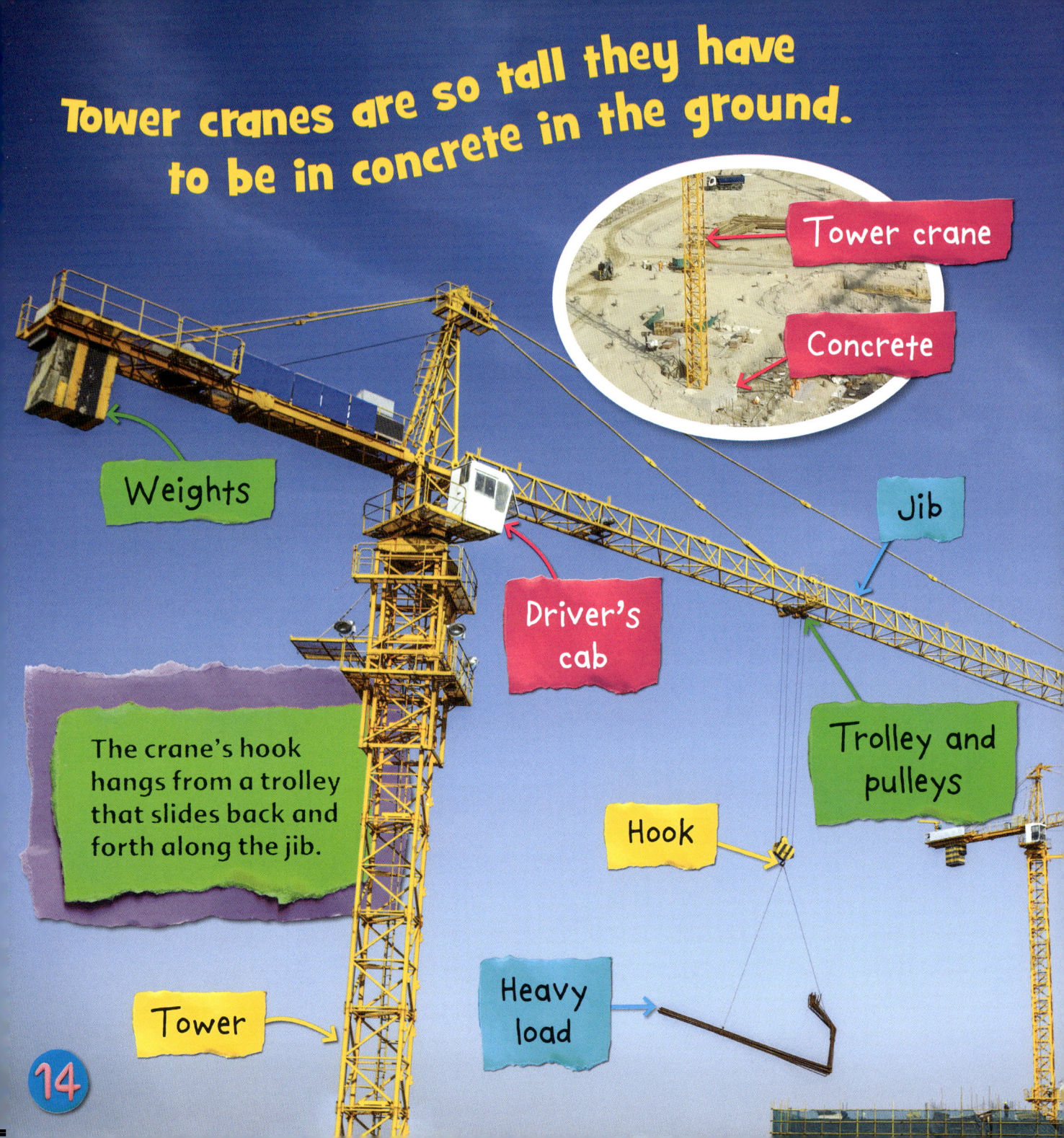

# Tower cranes are so tall they have to be in concrete in the ground.

Tower crane

Concrete

Weights

Jib

Driver's cab

The crane's hook hangs from a trolley that slides back and forth along the jib.

Trolley and pulleys

Hook

Tower

Heavy load

Crane driver

The crane driver climbs up metal ladders inside the tower to reach the cab.

The climb up can take 30 minutes!

The driver stays in the tiny cab for up to 12 hours.

They'd better remember their packed lunch!

How do they go to the loo?

They use a bottle or special bucket. Or they have to climb back down!

# A tower crane driver uses two joysticks in the cab to control the crane.

Inside a tower crane cab

**Left joystick:**
- Swings, or moves, the jib from side to side.
- Controls the trolley.

**Right joystick:**
- Moves the load up and down.

A crane driver knows how much weight their crane can carry. They never try to lift anything heavier.

A tower crane driver uses a walkie-talkie radio to talk to workers on the ground.

They may also talk to other nearby crane drivers to be sure their jibs don't touch!

Hook

A worker on the ground called a rigger attaches the crane's hook to the load.

# Crane drivers and their machines must work safely.

Pulley

A driver uses science to check all the parts of the crane regularly.

A crane driver will not use a crane that needs fixing. It could be dangerous!

A crane driver must always look down and up.

They must be sure that the crane's boom, jib or load does not hit electricity wires.

Crane driver

Radio

Binoculars

A tower crane driver may use binoculars to see what is happening on the ground.

The crane driver must be safe, too.

They wear protective clothes.

Hard hat

Headphones

Goggles

Hi-vis jacket

Gloves

Boots with metal toes

Crane drivers must keep watch on the weather.

If a storm with lightning happens, it can be dangerous to be in a crane.

All work must stop!

# Wind blows faster the higher up you are.

This crane was bent by strong winds in a bad storm.

A tower crane driver puts a wind vane machine on the jib.

It senses the wind's direction.

The wind vane turns the jib in the same direction as the wind. This stops the crane toppling over.

Wind vane

# The drivers of smaller cranes help build tall tower cranes.

Truck crane

Tower crane

This is Big Carl, the tallest crane in the world.

Big Carl is as tall as a skyscraper. It could lift 10 jumbo jets!

Weights

# Crane drivers use computers to give them information.

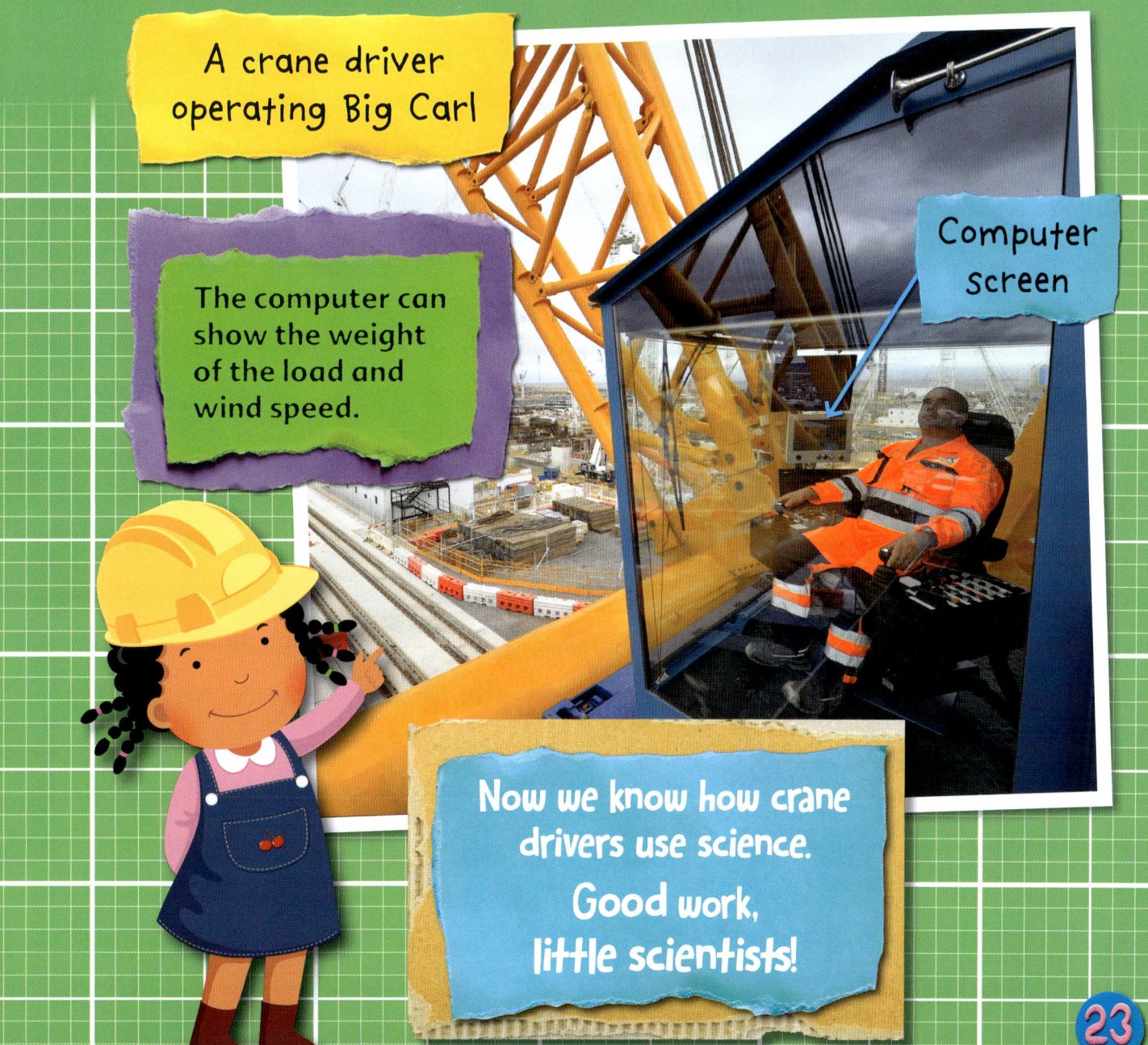

A crane driver operating Big Carl

The computer can show the weight of the load and wind speed.

Computer screen

Now we know how crane drivers use science.
Good work, little scientists!

23

# My Science Words

**boom**
The long arm of a crane that does the lifting. It has pulleys, cables and a hook.

**load**
The metal, concrete or other objects that a crane lifts, moves and lowers.

**concrete**
A thick, runny mixture of cement, sand, gravel and water. It turns rock hard as it dries.

**stabilise**
To position an object so it won't fall over.